This book is dedicated to all who find Nature not an adversary to conquer and destroy, but a storehouse of infinite knowledge and experience linking man to all things past and present. They know conserving the natural environment is essential to our future well-being.

THE STORY BEHIND THE SCENERY®

by Ronald A. Terry

Ronald A. Terry is a career employee of the National Park Service. A native of Missouri, Ron holds a master's degree from the University of Missouri. His career has taken him to Wind Cave, Death Valley, Kings Canyon, and Hot Springs National Parks, as well as several other NPS areas.

Wind Cave National Park, *in southwestern South Dakota, established in 1903, preserves unique cave formations and a free-roaming herd of bison.*

Front cover: Cave explorer in Southern Comfort Zone, photo by Arthur N. Palmer.
Inside front cover: Bison bull with prairie grasses; Page 1: Prairie dog family at burrow; photos by Kent and Donna Dannen. Page 2/3: Exploring near the Palette Room, photo by John P. Scheltens.

Edited by Cheri C. Madison. Book design by K. C. DenDooven.

WIND CAVE: THE STORY BEHIND THE SCENERY. © 1998 KC PUBLICATIONS, INC.
"The Story Behind the Scenery"; "in pictures... The Continuing Story"; the parallelogram forms and colors within are registered in the U.S. Patent and Trademark Office.
LC 98-65026. ISBN 0-88714-138-2.

Wind Cave, with its winding passageway leading into the dark unknown, remains one of the last frontiers to challenge the curiosity and imagination of mankind.

*W*ind Cave National Park encompasses two worlds—one visible, one hidden. A sea of waving prairie grasses covers about 70 percent of the park while the remainder is forested, with ponderosa pine being the dominant tree species. The prairie and forest support a diverse array of easily observable wildlife. Free-roaming bison present a window to the past when millions of their kind ruled the prairie realm. Eastern and western plants and animals existing side by side provide an unusual opportunity to experience the wealth of our nation's natural heritage.

Hidden beneath this scenic landscape lies Wind Cave. The extent and complexity of this mysterious underground realm defies comprehension. Over 80 miles of explored passages intertwine in a complex three-dimensional maze and exhibit the world's largest concentration of boxwork, an unusual cave formation. More than a century of exploration has revealed only a very small portion of the cave. The vast and unknown depths of Wind Cave represent one of the earth's last frontiers.

The natural struggle for dominance between pine forest and grasslands is altered by the invasion of exotic grass species such as smooth brome.

Wind Cave boasts the greatest variety and abundance of boxwork of any cave in the world. These rare and unusual honeycomb-shaped formations protruding from ceilings and walls are the signature formation of this cave.

The Hidden World

Wind Cave is named for the wind that often rushes through its 15-inch-diameter natural opening. The wind is caused by the natural process of air pressure equalization. When a low pressure system or storm moves into the area, air moves out of the cave from the higher pressure within. Conversely, when a high pressure system arrives in the area, air moves into the cave to the area of lower pressure within. The winds at the entrance have been recorded as high as 70 miles per hour, although winds of this magnitude are not common.

Wind Cave is one of the world's longest, oldest, and most complex caves. In 1997, the known length of the cave reached 80 miles. These 80 miles of passageways occur within the confines of one square mile on the surface. This is possible because the cave includes multiple levels interconnected by an incredible maze of intertwining passageways.

Common limestone cave formations, such as stalactites and stalagmites, are rarely seen at Wind Cave. Instead, the cave boasts the greatest variety and abundance of *boxwork* of any cave in the

◁ **The cave's constricted natural opening** concentrates air movement during times of changing air pressure. Even though the wind at the entrance may be very strong, air movement within the cave is barely noticeable due to the large volume of space through which the air must circulate. The rush of air through the natural opening led to the cave's discovery.

The impressive boxwork of Wind Cave is still developing at an imperceptible rate. As the softer, weathered rock surrounding veins of harder calcite continues to crumble, the calcite fins stand alone. The length of boxwork fins may range from a few inches to a few feet. This unusual and intricate cave formation appears throughout the middle and lower levels of the cave.

world. These rare and unusual honeycomb-shaped formations protruding from ceilings and walls are the signature formation of this cave.

Measurements of air movement at the cave entrance indicate that the 80 explored miles represent only about 5 percent of the actual length. An estimated 95 percent of the cave has never recorded the imprint of an explorer's boot nor had its features illuminated by a caver's lamp. Exploration continues in this vast underground wilderness where no one has ventured before.

CAVE FORMATION

Wind Cave's life story began during the Mississippian Period some 350 million years ago, when a warm shallow sea deposited sediments including the remains of marine animals. The cementation of these sediments formed rock layers of limestone and dolomite at least 300 feet thick. Collectively known as the Madison formation or Pahasapa Limestone, these rock layers are the home of Wind Cave.

The earliest cave passages are nearly as old as the limestone itself. The limestone deposited on the sea floor contained large amounts of intermingled gypsum—an easily dissolved, soft, white mineral, formed by the evaporation of sulfate-rich water. Chemical breakdown of the gypsum produced sulfuric acid, which dissolved small caves and pockets in the limestone. Dissolving and recrystallizing of the gypsum created cracks in the surrounding limestone, many of which refilled with gypsum. As freshwater seeped through the rock, this gypsum dissolved

△ **The Black Hills represent a classic example of a dome mountain. Younger layers of sedimentary** rock, covering the older granitic core, were arched downward and away from the center of the hills by the uplift of core. The sedimentary rock layers in the area of Wind Cave National Park are tilted an average of five degrees to the southeast. Time and the elements worked their erosional magic on these soluble sedimentary rocks, exposing the granitic core of the hills and creating concentric rings of exposed sedimentary rocks. The Pahasapa Limestone which contains Wind Cave is exposed as sharp, nearly vertical cliffs when eroded by the relentless action of surface streams.

The serene landscape of rolling hills masks the existence of the subterranean world hidden below. Surface streams provide evidence that the underground geologic structure could harbor a cave system. Streams enter the park but sink into the ground before they exit. Miles away, the water emerges in a spring after flowing through porous rock layers that also contain the vast underground labyrinth of Wind Cave.

and was replaced with calcite. The small openings produced by these natural processes represent the birth of Wind Cave.

For about 300 million years, a series of seas alternately covered the land and then receded. When the land was exposed, freshwater found its way into the cave dissolving the rock and enlarging the existing cave passageways. Sediments from these later seas washed into the cave creating the reddish-brown paleofills seen along cave tour routes today.

Cave formation accelerated when the Black Hills were uplifted about 60 million years ago. This uplift caused existing cracks in the sedimentary rocks to enlarge and new cracks to form. Underground water was then able to move more freely through the limestone but was unable to drain out. This allowed standing acidic water to dissolve the limestone and dolomite uniformly along the existing fractures, enlarging the cave. Some 40 to 50 million years ago outlets for the water developed and the cave drained. Today, the water table is about 500 feet below the surface at the Lakes, where the slow process of dissolving and enlarging the cave continues.

Speleothems

Speleothems, often called cave formations or features, decorate the interior of the cave. *Boxwork*, a rare type of speleothem, occurs in great profusion in Wind Cave, especially in the dolomite of the middle and lower levels of the cave. These thin fins of calcite, resembling honeycombs or open-ended boxes, protrude from cave walls and ceilings. Boxwork is composed of calcite formed in cracks by the breakdown of gypsum before the cave existed. Naturally produced hydrogen sulfide and sulfuric acid caused the dolomite between the calcite veins to be crumbly, allowing water to move easily and dissolve the weathered rock from between and around the boxwork fins, leaving them standing out in relief. Some of the best displays along tour routes are in the Post Office, the Elks Room, and the Temple.

Most speleothems are formed by the action of water seeping into the cave from the surface. The water picks up carbon dioxide, forming a weak carbonic acid. This acidic water dissolves

◁ **Frostwork is** composed of delicate, needle-like crystals of aragonite, a carbonate mineral related to calcite. Frostwork grows on many different types of cave surfaces, including other cave features such as this cave popcorn. In Wind Cave, frostwork tends to occur in areas that experience above-average air flow, perhaps indicating that more rapid evaporation of seeping water is an important factor in causing the crystals to form as they do.

the rock and picks up calcium carbonate as it passes through. Emerging from the rock, the water loses carbon dioxide to the cave air, reducing its capacity to hold the dissolved calcium carbonate. The calcium carbonate then precipitates on the walls and ceilings of the cave in the form of calcite crystals that form a variety of different speleothems.

Common calcite crystal decorations in Wind Cave include *cave popcorn*, rounded, calcite nodules; *frostwork*, a delicate, feathery formation sometimes several inches in length; *helictite bushes*, resembling gnarled tree branches; and *dogtooth spar*, consisting of angular, calcite crystals resembling teeth. Paper-thin sheets of calcite known as *calcite rafts* float on some of the cave's lakes. Flowstone and dripstone formations such as

◁ **Cave passages formed in the last 60 million** years sometimes intersect older passages filled with a reddish-brown conglomerate. The conglomerate is composed of sand, clay, and other sediments that washed into the older cave during the Pennsylvanian Period.

Bats rarely enter Wind Cave but are frequently observed flying erratically above the cave as they search for insects on a warm summer evening. Since they do not roost in the cave, they seek out sheltered spots in trees, rock cliffs, and buildings. The big brown bat is but one of six species known to exist in the park.

stalactites, *stalagmites*, and *draperies* exist beneath areas of frequent surface water but are rarely seen. Fossils of brachiopods, sponges, corals, and other marine life are most often found in the upper levels of the cave.

CAVE LIFE

Wind Cave is an example of an extremely simple ecosystem. Energy needed to sustain life is almost nonexistent. Darkness and cool temperatures prevail, water exists in very small quantities except in the tranquil lakes deep within the cave, and food in the form of organic material is very scarce. Even with these conditions, some life manages to survive here.

Lights used to illuminate the cave tour routes have caused green and blue-green algae to grow on rock surfaces nearby. The National Park Service is constantly working to control and eliminate this unnatural plant growth. No other types of plant life are known to exist in the cave.

Bats are rarely found in Wind Cave. Six species of bats have been seen in the cave, with the little brown myotis being the most common. Winds, known to reach speeds of 70 miles per hour through the natural opening, may act as a barrier to their movement. Bats are not known to hibernate in the cave. The constant 53-degree temperature is believed to be too warm for hibernating bats. If their body temperature remains too high during hibernation, fat reserves which must support them are burned off too quickly and they die.

Infrequent visitors to the cave include wood rats, tiger salamanders, and an occasional errant snake. They do not venture far into the cave and could not survive its cool and food-barren depths. Crickets, beetles, and spiders are also known to dwell just inside the cave entrance. In the depths of the cave, small insects known as springtails live in damp areas and feed on microorganisms and decaying organic matter brought in by surface water. At least six types of springtails have been identified, and others may still be discovered.

The cave's simple ecosystem is of special interest to scientists studying more complex ecosystems and their interrelationships. A simple cave system with few variables is an excellent place to test hypotheses relating to more complex systems. Caves such as Wind Cave may hold the key to better understanding the complex web of life in which we all live.

SUGGESTED READING

HARRIS, ANN G., ESTHER TUTTLE, and SHERWOOD D. TUTTLE. *Geology of National Parks.* Dubuque, Iowa: Kendall/Hunt Publishing Company, 1996.

MOORE, G. W., and G. N. SULLIVAN. *Speleology, Caves and the Cave Environment.* St. Louis, Missouri: Cave Books, 1997.

PALMER, ARTHUR N. *Wind Cave, An Ancient World Beneath the Hills.* Hot Springs, South Dakota: Black Hills Parks and Forests Association, 1995.

Cave popcorn is one of the most ◁ common formations in Wind Cave. These calcite nodules are the result of small amounts of water seeping uniformly through limestone, depositing calcium carbonate as the water evaporates. The nodules can also form where dripping water splashes on the cave wall, coating it with a thin film of moisture. Often occurring in large clumps, the nodules resemble popcorn—thus, the name.

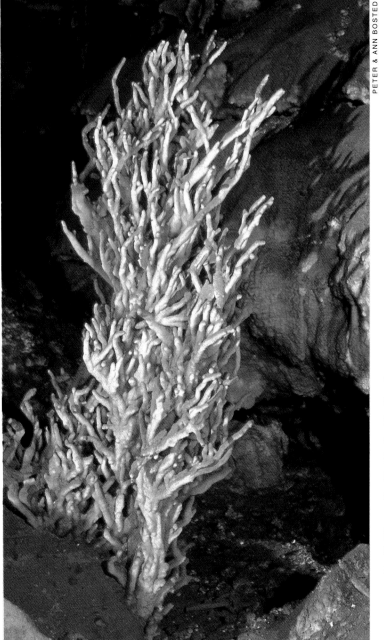

◁ **Helictite bushes are composed of** calcium carbonate formed when minute amounts of water seeping from the limestone move by capillary action upward through the formation's central tubes. The calcium carbonate is then deposited on the tip of the branches causing them to grow longer. Their delicate upward growth seems to defy gravity. The largest helictites occur near the Lakes, and the largest known specimen is an astonishing five feet tall.

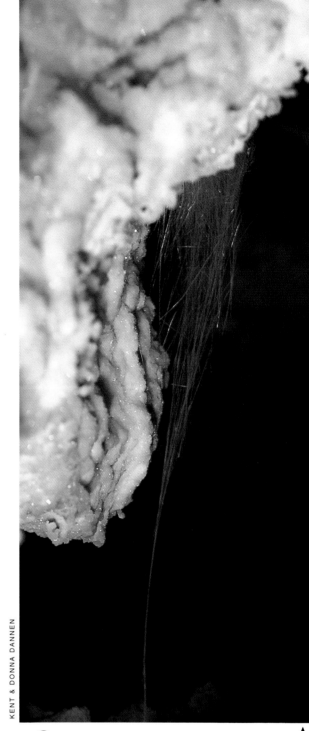

Gypsum, a white mineral, appears in ▷ the cave in several forms ranging from flower-like clusters on the surface of rocks to long, wispy, strands resembling hair, called gypsum "beards." These extremely delicate formations will move with the slightest movement of air.

◁ **Soda straws are a type of stalactite** characterized by thin, hollow tubes hanging from the ceiling. The formations are, in fact, stalactites in their infancy. Given enough time, these fragile tubes will grow and become a solid stalactite when the inner tube is clogged by mineral deposition and water continues to drip down the outside of the tube.

△ **Preserving cave resources is a primary concern for cavers traveling through known** areas of the cave as well as newly discovered areas. Fragile formations, such as the carpet of calcite crystals on the floor of a cave room, must be carefully avoided to prevent irreparable damage.

◁ **Chert layers in** the middle level of the cave often form flat ceilings and contribute to the formation of unusual cave features such as this upside-down "mushroom." The surrounding, more-soluble rock dissolved or weathered away, leaving the more-resistant chert in place.

Passageways in the lower levels of the cave are often high, narrow fissures, following fractures in the bedrock.

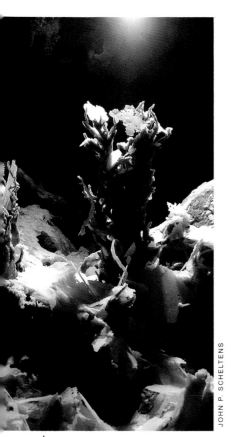

Calcite rafts, once floating on the surface of water, now decorate and surround a helictite bush in the depths of the cave.

Santa's Frosted Forest is named for aragonite formations resembling trees in a snow-covered landscape. Leaning trees may be the result of air flow causing the crystal trees to grow at an angle.

◁ **D**ripping water deposits calcite that accumulates to form dripstone formations, including stalactites which hang from the ceiling, stalagmites which build up from the floor, and columns which form when stalactites and stalagmites meet. Wind Cave is a relatively dry cave and, therefore, does not produce large amounts of dripstone.

In some places ▷ in the cave, patterns resembling fossil ferns appear on chert surfaces. These features, called manganese dendrites, are actually dark crystals of manganese oxide formed along thin cracks in the bedrock. The crystals emanate from the mineral source forming a branching linear pattern.

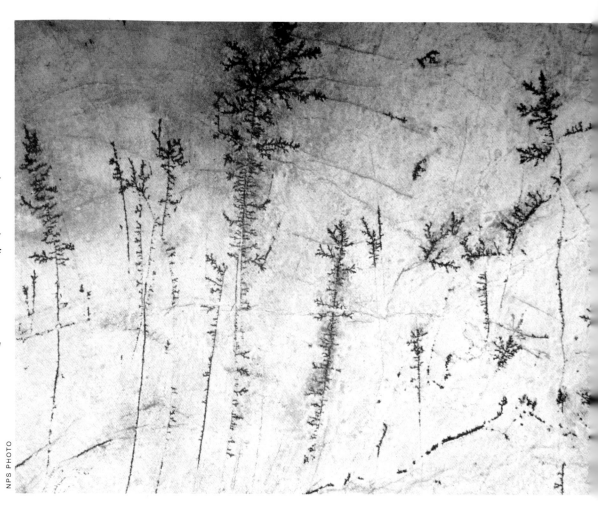

△ **Gypsum starbursts are** patterns found on the surface of weathered bedrock powder. Starbursts are formed as water carrying dissolved gypsum evaporates, leaving crystallized gypsum behind.

Aragonite △ frostwork often forms lacy clusters of crystals up to several inches in length.

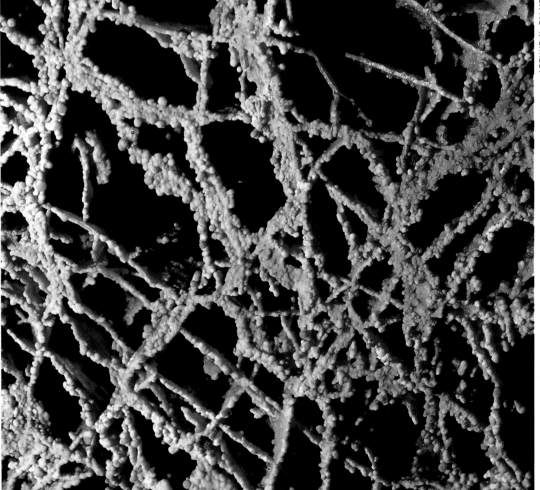

◁ **Boxwork fins** are often frosted with cave popcorn nodules formed when the boxwork is exposed to small amounts of evaporating water.

*The wildlife grandeur of the past
was best symbolized by the bison
whose numbers are believed to have reached 60 million.
The 350 bison roaming the park's grasslands
today provide a visual reminder
that these great beasts were nearly extinct.*

Life on the Edge

Wind Cave National Park is located where prairie grasslands blend with the forested slopes of the Black Hills and both eastern and western plant and animal species meet at the edge of their ranges. This set of conditions allows for an uncommon diversity of life. Three distinct plant communities—the prairie, the forest, and woody ravines—blend together to provide habitats for a diverse mixture of wildlife species. This mosaic of life gives the visible world of the park its exceptional variety and appeal.

The Prairie

Wind Cave's mixed grass prairie includes plant species from both the more arid, western, shortgrass prairie and the moister, eastern, tallgrass prairie. Midgrass species such as western wheatgrass, needle and thread, and little bluestem mingle with tallgrasses (including big bluestem and switch grass) as well as shortgrasses such as buffalo grass and blue grama. The mixture of grass species varies from year to year depending upon the amount of available moisture. Grasses account for about 20 percent of prairie plant species but provide about 80 percent of the ground cover.

Abundant wildflowers such as black-eyed Susans, sego lilies, woolly verbena, western wallflowers, purple coneflowers, and upright prairie

◁ **Widely spaced Rocky Mountain juniper and** ponderosa pine are evidence of the slow advance of the forest. Seeds from these individuals, carried by the wind, a bird, or water, can find receptive soil, and a new generation will advance farther into the grassland.

△ **Bison calves, born in April and May, are** more reddish in color than the adults. This reddish coat will darken before they are one year old. The bond between cow and calf is strong, and cows have been known to charge a person who approaches too close to a calf.

coneflowers sprinkle the prairie with color. On drier southern slopes, small soapweed send their flowering central spikes skyward.

Historically, the prairie nourished and supported an assemblage of wildlife whose numbers defy the imagination. The wildlife grandeur of the past was best symbolized by the bison whose numbers are believed to have reached 60 million. The 350 bison roaming the park's grasslands today provide a visual reminder that these great beasts were nearly extinct.

Bison are often called buffalo, but this is a misnomer since true buffalo are native to Asia and Africa. Bison have massive heads and a distinct shoulder hump. They can weigh as much as 2,000 pounds. While they often appear to be slow and docile, in fact they can run faster than a person and their temperament is unpredictable. They will not tolerate the approach of a person, so the only safe way to enjoy bison is from a distance or the relative safety of an automobile.

Pronghorn share the grasslands with bison. This strikingly handsome creature, often called an antelope, is North America's fastest mammal. These graceful animals can glide across the prairie at speeds up to 60 miles per hour. Mule deer with their characteristic large ears and bounding gait are most often seen in small groups on the prairie near the forest edge.

Black-tailed prairie dogs are large burrowing rodents whose antics are always entertaining. Eradicated from much of their former range, they are protected and thriving here. Prairie dogs live in colonies commonly called "prairie dog towns" which often cover many acres. These towns attract a wide variety of other wildlife including coyotes,

▲ **The song-like bark of a single coyote** on a still prairie evening often grows to a chorus as others join in. These efficient predators, found throughout the park, prey primarily on small animals such as mice, rabbits, and prairie dogs, although they have been known to successfully hunt weak or injured deer. Watch for them prowling prairie dog towns.

who are often seen cruising the towns looking for a meal. Badgers, golden eagles, and hawks hunt here as well. Prairie rattlesnakes and burrowing owls make their homes in prairie dog burrows. Bison are attracted to the new growth of grasses caused by the prairie dogs' habit of cropping grasses low to provide a view of approaching predators. They also seek relief from insect pests by wallowing in the exposed dirt of the towns.

A variety of birds are associated with the prairie but none more so than the western meadowlark whose melodious song could easily be described as the prairie anthem. Other common birds include the common nighthawk, vesper sparrow, upland sandpiper, and mourning dove. The prairie is also the hunting realm of red-tailed hawks and the smaller and more colorful American kestrel.

PONDEROSA PINE FOREST

Ponderosa pine reaches its easternmost range in the Black Hills. Almost one third of the landscape of Wind Cave National Park is ponderosa pine forest. Ponderosa pine are well adapted to the dry slopes of the park where the average annual precipitation is 17 inches. Deep roots anchor it against strong winds and allow it to grow well in rocky soils. Low-growing common juniper form a mat in patches on the forest floor, providing shelter for small ground-dwelling animals. Early blooming pasqueflowers

◀ **Elk are normally** shy and secretive, preferring to stay hidden in the protective cover of the forest. Venturing into meadows to feed, they will quickly escape back into the forest at the slightest sign of danger.

and star lilies dot the open forest floor, and shooting stars display their torch-like blooms.

Eastern elk, former inhabitants of the park's rolling hills, are now extinct and have been replaced by Rocky Mountain elk. Resting in the protection of the pine forest during the day, elk venture into open meadows to feed at dawn and dusk. About 350 of these magnificent animals reside in the park, split among three distinct herds. The shrill, haunting call of a bull elk during the mating season in September and October is a wild sound never to be forgotten.

A few white-tailed deer seek shelter among the pines. The forest calm is broken by the chatter of red squirrels, least chipmunks scurry about the forest floor, and porcupine feed on the soft inner bark of pines. Forest predators include an occasional bobcat or mountain lion, but the bears and wolves are gone. The forest provides suitable habitat for a variety of birds including red crossbills, white- and red-breasted nuthatches, black-capped chickadees, colorful western tanagers, and great horned owls. Hairy woodpeckers and northern flickers search for insects among the pines.

Woody Ravines

The park's three small streams—Beaver Creek, Highland Creek, and Cold Springs Creek—and numerous intermittent watercourses in small ravines sustain a plant community of eastern hardwood trees and woody shrubs. Bur

△ **Porcupines, large and primarily nocturnal** rodents, make their home in the forested areas of the park. They are often found in trees, where they dine on twigs, leaves, and the inner bark of branches. A bristling coat of sharp, hollow quills provides a formidable and effective defense against predators.

◁ **Mule deer, often** seen munching on grasses and shrubs, also frequent natural mineral licks used by bison. Mule deer are far more numerous than white-tailed deer in the park. They remain wary of coyotes and an occasional mountain lion, their only remaining predators.

Acrobatic white-breasted nuthatches ▷
creep headfirst down tree trunks while
searching for insects beneath the bark. As winter
approaches, they store food under the bark.

◁ **S**piderwort blooms in ravines and
moist soils in June and July, closing
its blooms during the heat of the day.

oak, American elm, box elder, green ash, and plains cottonwood spread their leafy branches. Chokecherry, American plum, golden currant, and western snowberry provide sheltering thickets for eastern cottontail rabbits and other small mammals and a food supply for birds and other animals. American goldfinch, yellow warblers, rufous-sided towhee, western wood peewees, and many other bird species add their melodies to the prairie wind. Wildflowers, including wild roses and sweet-smelling mints, take advantage of the moister soils found here. Poison ivy also can be

◁ **B**lack-capped chickadees, active birds
of the thickets and forest edge, are often
seen in groups in the lower branches of trees.

Prescribed fires are used under carefully monitored conditions to simulate the role of natural fire in the park. By removing dense "doghair" stands of young ponderosa pines, the natural balance between forest and prairie is maintained. Fire benefits the prairie by rejuvenating prairie grasses, and it is proving to be an important tool for preserving native prairie by controlling and destroying some exotic plant species.

found here in abundance, so hikers should be cautious. The woody ravines connect the forested hills and the open prairie with ribbons of more luxuriant plant growth.

Preserving the Balance

The park's native grasslands and the wildlife that depend upon them are being threatened by an invasion of non-native or exotic plant species. At least 46 exotic plant species, including Canada thistle and smooth brome, are known to exist in the park. These plants, with little or no value to the park's wildlife, are spreading and replacing native species. The control and eradication of these invaders is now a major focus of the park's resource management program.

The prairie is also threatened by encroachment of the ponderosa pine forest unless fire is present to maintain the balance. The National Park Service has used fire under controlled conditions at Wind Cave since the 1970s to simulate the role of natural fires in maintaining the balance between forest and prairie.

Similarly, it is important to preserve the balance between the cave and the surface world. Today these relationships are better understood. We are more aware that past actions are causing impacts on the cave today. The introduction of electric lights caused the growth of algae. The opening of artificial entrances changed air flow and the cave environment. Millions of cave visitors inevitably affect the cave's fragile ecosystem. The National Park Service is working to restore the cave's natural conditions and to prevent additional man-caused impacts.

SUGGESTED READING

BROWN, LAUREN. *Grasslands.* The Audubon Society Nature Guides. New York: Alfred A. Knopf, 1985.

COSTELLO, DAVID F. *The Prairie World.* New York: Thomas Y. Crowell Company, 1969.

FROILAND, SVEN G. *Natural History of the Black Hills and Badlands.* Sioux Falls, South Dakota: The Center for Western Studies, 1990.

SAMPLE, MICHAEL S. *Bison, Symbol of the West.* Helena and Billings, Montana: Falcon Press Publishing Co., Inc., 1991.

VAN BRUGGEN, THEODORE. *Wildflowers, Grasses, & Other Plants of the Northern Plains and Black Hills.* 4th ed. Interior, South Dakota: Badlands Natural History Association, 1992.

WHITNEY, STEPHEN. *Western Forests.* The Audubon Society Nature Guides. New York: Alfred A. Knopf, 1985.

Overleaf: Red Valley derives its name from the colorful shale of the Spearfish formation. Photo by Clint Farlinger.

△ **M**ature bison bulls return to the herds during the rut or mating season in July and August. Bulls are short-tempered at this time and even more dangerous than usual. Their aggressive behavior includes lip curling and a deep, guttural bellowing that can be heard for long distances on the prairie.

Dusting and rolling in the dirt are △ ▷ common bison activities. They create depressions in the dirt called "wallows" where they roll to aid in removing unwanted hair and to coat themselves with dirt to provide temporary relief and protection from insect pests. Large shoulder humps prevent them from rolling completely over. Prairie dog towns also serve as places to stir up dust by using their hooves or horns. During the rut, agitated bulls will exhibit their aggression by horning the ground.

Pronghorn are not closely related to any other living animal. Their keen eyesight and surprising speed serve them well on the open grasslands. Nourished by leafy shrubs and small broad-leaved plants, they ▽ co-exist easily with bison, who depend upon prairie grasses.

◁ △ **With the arrival of warm spring** weather, bison begin to shed their thick winter coats. To aid in removing the hair, bison often use trees as rubbing posts, scarring them in the process. Tufts of discarded hair become a treasure for nesting birds.

Black-tailed prairie dogs are large rodents belonging to the squirrel family. They were named "prairie dog" due to the similarity of their call to the bark of a dog. They live in burrows in colonies or "towns" which may reach several hundred acres in size. Prairie dog towns are the community centers of the prairie, attracting many wildlife species either for food or shelter. Prairie dogs keep plants clipped short, allowing them to have an open view of approaching predators. As the grasses attempt to grow back, bison are attracted to the new and tender shoots. Forbs growing in the newly exposed soil provide an attractive food source for pronghorn. Prairie dogs provide an abundant food source for predators such as eagles, hawks, owls, coyotes, and badgers. Prairie dog burrows become homes for burrowing owls, rabbits, rattlesnakes, and black widow spiders.

△ **G**olden eagles swoop down upon unsuspecting prairie dogs who stray too far from the safety of their burrows.

Badgers are tenacious ▷ predators capable of digging prairie dogs out of their burrows. These powerful members of the weasel family are solitary hunters of the open spaces.

Burrowing owls survey the landscape from the mound of an abandoned prairie dog burrow where they live. These small ground-dwelling owls feed on insects, small rodents including young ▽ prairie dogs, and snakes.

Prairie Grasses

One third of the continent was once a sea of grass. Prairie grasses historically sustained a diverse assemblage of wildlife, providing both food and shelter. Grasses are also important for human survival since many of our most important food crops, such as wheat, corn, rice, oats, barley, and rye, are grasses. Meat-producing animals depend upon grasses for survival and growth.

Today, unspoiled grasslands are disappearing. Much of the nation's prairie expanse has succumbed to the plow and to overgrazing. The natural prairie ecosystem is also threatened by the unintentional introduction of non-native plant species. Native grasses are being replaced with these exotic plants, that often grow in monocultures, ending the diversity of plants so important for a healthy prairie ecosystem. The suppression of natural fires has also played a role in hindering the natural revitalization of prairie grasses. Places like Wind Cave National Park, attempting to preserve 20,000 acres of grassland in its natural condition, are important islands of hope for the survival of our prairie heritage.

△ **Big bluestem,** an eastern tallgrass species, can reach six feet in height. Commonly growing in lower, moister locations, big bluestem is an excellent forage grass for wildlife.

◁ **Side oats grama and** western wheatgrass are medium-height cool-weather grasses characteristic of the mixed grass prairie. Side oats grama is named for its oat-like "seeds" hanging in two rows.

Western wheatgrass ▷ can be identified by its blue waxy appearance.

△ **The dominant grass of the western shortgrass prairie is blue grama,** which only reaches a few inches in height. Hot temperatures and little moisture are conditions well tolerated by this hardy species.

One-sided seed △ spikes characterize blue grama.

Red **Valley, on the eastern side of the park, contains an expanse of prairie grassland that often attracts** herds of bison. A variety of grasses and forbs and a few small springs provide the nourishment and water necessary for numerous prairie wildlife species. Boland Ridge, bordering the valley on the ▽ east, separates the Black Hills from the Great Plains extending to the horizon and beyond.

*The first recorded discovery of the cave
occurred in 1881 when two brothers,
Tom and Jesse Bingham, were hunting in the area.
The sound of rushing air
led them to the natural opening.*

Those Who Came Before...

Knowledge of the earliest people to inhabit the region is sketchy at best. Evidence does indicate that as early as 6,000 years ago, during the Early Archaic period, small groups of nomadic hunters and gatherers were present here. The hills provided shelter from the more severe weather conditions of the surrounding plains, stone for the manufacture of tools and weapons, wild game, and plant foods.

Later, American Indian tribes including the Kiowa, Crow, Cheyenne, Ponca, and Lakota (Sioux) were attracted to the Black Hills and claimed them as their domain. A long series of conflicts among these groups ended with the Lakota gaining dominance over the hills. With the exception of scattered stone tipi rings, little evidence remains to record their presence. Tipi rings near Wind Cave make it seem likely that the location of the cave entrance was known. A Lakota story tells of bison emerging from a cave in the Black Hills to populate the great plains. The cave may have been Wind Cave—or perhaps another of the dozens of caves in the hills.

Cave Development

The first recorded discovery of the cave occurred in 1881 when two brothers, Tom and Jesse Bingham, were hunting in the area. The sound of rushing air led them to the natural opening where the force of the wind blew Tom's hat off, or

△ **Several people claimed the discovery of Wind Cave.** The most widely accepted version credits the first recorded discovery to Tom and Jesse Bingham. Some sources claim that Jesse actually discovered the cave opening, but because his reputation was tarnished, his brother Tom was ultimately credited with the discovery. There is no evidence that either of the brothers ever entered the cave.

▲ **Wind Cave was first developed as a tourist attraction by Jesse McDonald and his family. The McDonalds** lived in a log house constructed at the cave entrance. To allow easier access to the cave, a vertical entrance was excavated large enough to accommodate visitors entering and exiting by a ladder. To shelter the entrance, a small structure was built over it. These improvements, coupled with work completed in the cave to enlarge cave passages, attracted more visitors to the cave.

so the popular story goes. A local newspaper article claimed that Charlie Crary of Custer, South Dakota, entered the cave in the fall of 1881. The cave remained a local curiosity until 1890, when the South Dakota Mining Company filed a claim to be able to extract minerals. Jesse McDonald was hired to manage the mining claim but was later discharged when it became apparent that minerals of commercial value and quantity were not present in the cave.

McDonald remained and began developing the cave for tours. Structures were constructed above the cave, and passageways were enlarged by blasting. McDonald's two sons, Alvin and Elmer, explored the cave using candles and balls of twine. Alvin was enthralled with the cave and faithfully maintained a diary and a map recording their discoveries. An entry in Alvin's diary in 1891 stated "...have given up the idea of finding the end of Wind Cave." This would prove to be a very prophetic statement. The McDonalds survived by charging for cave tours and extracting cave specimens which they sold to private collectors and dealers.

Jesse McDonald enlisted John Stabler of Hot Springs, South Dakota, as a partner, and they formed the Wonderful Wind Cave Improvement Company to develop and promote the cave as a tourist attraction. John Stabler was to receive a share of the profits from the cave and all the proceeds from a hotel he built near the cave entrance. The cave tour business increased with Stabler's promotional efforts. In 1893, Jesse and Alvin McDonald attended the Columbian Exposition in Chicago, selling cave specimens and promoting the cave. Alvin was ill with typhoid fever when they returned to Wind Cave and died shortly thereafter at the age of 20.

A dispute developed between the McDonalds and the Stablers over the division of the receipts

The Partners

The McDonald family, headed by Jesse McDonald, operated the cave as a commercial enterprise combining tourism and mining of cave specimens. The youngest son, Alvin, was considered to be the earliest cave explorer. He spent countless hours underground exploring new leads, guiding others, improving cave passages, and blasting and removing cave specimens to be sold. Dead at the young age of twenty, his grave overlooks the entrance to Wind Cave.

John Stabler was the promoter in the Wind Cave partnership. He claimed to have found a "petrified man" near the cave. He exhibited this curiosity in the cave and charged a fee to view it. Stabler also brought a mind reader to the cave for the purpose of finding a lady's scarf pin hidden in the cave. After much publicity for this stunt, the pin was reported to have been found. These antics served to attract more attention to the cave.

from their business. The situation was further complicated by a suit filed against the Wonderful Wind Cave Improvement Company by the South Dakota Mining Company in an attempt to remove the two families from the site. A complicated series of suits and countersuits followed.

The General Land Office had the task of settling the dispute over land and cave ownership. It ruled that a homestead claim filed by Jesse McDonald on the land surrounding the cave entrance was void since the requirements of the Homestead Act were not fulfilled. It further ruled that the lands were non-mineral and, therefore, the existing mining claims of the other litigants were also voided.

The Commissioner recommended that the land be set aside as a "public reserve," and the lands around Wind Cave were permanently withdrawn from settlement in 1901. Meanwhile, cave tours continued under the authority of the Department of the Interior. In 1902, legislation to make Wind Cave a national park was introduced and passed in both houses of Congress. On January 9, 1903, President Theodore Roosevelt signed the legislation establishing Wind Cave National Park as the nation's seventh national park, the first devoted to a cave.

William A. Rankin, the first superintendent, and the superintendents who followed him attempted but failed to convince the Department that the park should receive funding to repair and improve its facilities. There were proposals to deauthorize it as a national park or to change it to a national monument as "more impressive caves" were discovered. Strangely enough, bison were responsible for ending the debate about whether or not Wind Cave should remain a national park.

THE NATIONAL GAME PRESERVE

By 1900 only one wild herd of bison existed in this country, in Yellowstone National Park. Groups such as the American Bison Society sought to restore bison to their former ranges. The Society selected Wind Cave National Park as a suitable place to reestablish a herd, and Congress passed a bill in 1912 establishing the Wind Cave National Game Preserve under the management of the U.S. Biological Survey. Fourteen bison came from the New York Zoological Society in 1913. Additional bison were transferred from Yellowstone National Park, and by 1918 the herd had increased to 42 animals. Elk and pronghorn were also reintroduced to the new preserve.

△ *Today, bison herds roam freely throughout* the park. When reintroduced to the newly established national game preserve in 1913, bison were restricted to about 4,500 acres including a small fenced pasture near the main road. The high visibility of the animals sparked pubic interest and curiosity, causing park visitation to increase tenfold from 1913 to 1922. This increase in visitation allowed the National Park Service to justify funding for the first major facility improvements for the new park.

▲ **Beaver Creek Canyon is spanned by an impressive concrete arch bridge, listed on the National** Register of Historic Places. This bridge, with symmetrically curving ends, was completed in 1929 as part of a larger effort to improve roads and promote tourism throughout the Black Hills. Cement for the arch ribs was poured continuously day and night until completed.

In 1935, management of the game preserve was transferred to the National Park Service. The game preserve not only played a role in saving bison from extinction as a species but also attracted more visitors to the park. The addition of wildlife was a crucial factor in retaining the park's status as a national park.

THE CIVILIAN CONSERVATION CORPS

As visitation to the park increased in the 1930s, so did funding for construction of new facilities and operation of the park. The park received a needed boost in 1934 when a Civilian Conservation Corps Camp (NP-1) was established at Wind Cave. Company 2754 constructed buildings, fences, roads, trails, wildlife facilities, bridges, retaining walls, road culverts, a new cave entrance, cave stairs and trails, and a new electrical lighting system for the cave.

The visitor center and most of the park's buildings are designated a National Historic District preserving examples of the rustic architecture of the Civilian Conservation Corps. Their most impressive single accomplishment was the construction of the elevator building and double elevator shaft to provide better public access to the cave. Still in use today, these sturdy structures stand as a tribute to the efforts and skills of the approximately 2,000 young men who toiled here from 1934 to 1939.

THE NATIONAL PARK SERVICE ROLE

The National Park Service attempts to accommodate an increasing number of visitors while protecting the natural and cultural resources which make the park a special place. These dual purposes can be challenging to accomplish without one compromising the other. The efforts of the National Park Service build upon the accomplishments of all those who came before for the benefit of those who will come after.

SUGGESTED READING

NATIONAL PARK SERVICE. *Wind Cave National Park, South Dakota.* National Park Service Handbook. Washington, D.C.: U.S. Government Printing Office, 1979.

GILDART, BERT, and JANE GILDART. *Hiking South Dakota's Black Hills Country.* Helena and Billings, Montana: Falcon Press Publishing Co., Inc., 1996.

ROGERS, HIRAM. *Exploring the Black Hills and Badlands.* Boulder, Colorado: Johnson Books, 1993.

△ **The Civilian Conservation Corps played a major** role in developing the park for both visitors and employees. Involved in the construction of the visitor center, maintenance buildings, cave lighting, fencing, road improvements, trails, and bridges, their crowning accomplishment was the construction of the elevator building completed in 1938, using native stone and timbers. This handsome building still serves the same purpose today.

Gathering storm clouds on a summer afternoon bring the promise of rain to nourish prairie plants and continue the age-old cycle of life on the prairie. ▽

Wildflowers at Wind Cave

△ **T**he blooms of prairie golden pea brighten the prairie and open forest in May and June.

Grasses and broad-leaved plants △ with showy blooms, called forbs, combine to form the dense biomass of a mixed grass prairie. The composition of plant species will change with the season and climatic conditions.

◁ **W**ith a blooming season extending into September, woolly verbena can still be in bloom when the first frost covers the prairie. Most prairie forbs complete their blooming before the arrival of frost. Repeated frosts will cause verbena and other perennial prairie plants to transfer their energy into their root systems. The aboveground structure of the plant will die during the winter months, but the roots will be alive, although dormant. New growth reappears with the warmth of spring.

△ **The arid nature of the prairie is exemplified** by the occurrence of plains prickly-pear cactus. The waxy outer coating of the plant helps to conserve water stored in its succulent pods.

Open areas with disturbed △ soils often host curlycup gumweed, which exudes a sticky substance responsible for the plant's name.

◁ **Streamsides, forested slopes,** and meadows where the soil is moist and deep support clumps of western blue flag, also known as Rocky Mountain iris. Blooming from May to July, these native, wild irises closely resemble their domesticated cousins commonly grown in household flower gardens.

△ ***Cave exploration conducted by volunteer caving groups or "grottos" continues under*** the direction of the National Park Service. In the 1990s alone, exploration efforts revealed over 27 new miles of cave. The practice of exploring passages that have noticeable air flow has led to discoveries of some of the largest rooms and passages known to exist in Wind Cave. Caving is dirty, grueling work but it rewards explorers with the opportunity to see places never before seen by mankind, such as this passageway named Gypsy Road.

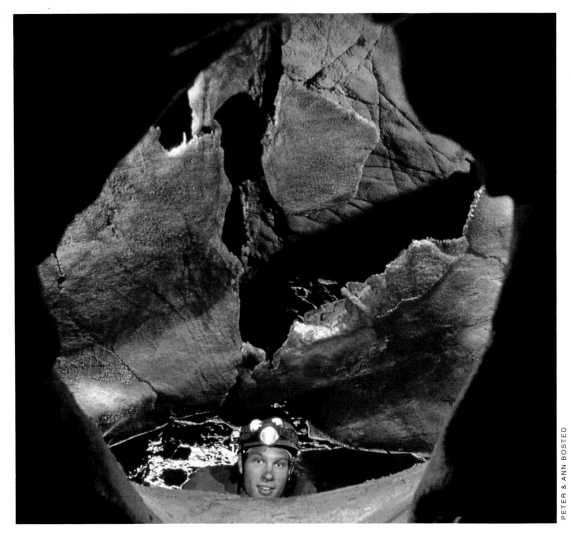

◁ **P**eering into the dark recesses, a caver contemplates what may be waiting beyond the light. Many cave discoveries have occurred only after long and arduous crawls through spaces so tight that both the caver's chest and back are touching the rock. Exploration teams include at least three well-equipped and well-trained cavers.

Early cave ▷ tours were more primitive than those of today. A park ranger relates the adventure of touring the cave by the flickering light provided by candle buckets. Used in the early 1900s to provide directional lighting, this apparatus served as an early-day flashlight. Today, special candlelight tours are offered during the summer months.

△ **An extensive boxwork display covers the walls and** ceiling in the Temple, a room on the Fairgrounds tour route. With numerous crevices and openings visible from the trail, this room also provides a good example of the complex maze of passageways that characterize Wind Cave.

△ **Angular calcite crystals with sharp** tips resembling teeth are known as dogtooth spar. These crystals are found most often in pockets in the limestone.

Aragonite crystals occur in several forms, ▷ including a Christmas-tree shape like this one in Christmas Tree Park. Reaching a foot or more in height, the white crystals may be hidden by the fine dust of weathering limestone bedrock.

▲ **Fossil coral provides evidence of the marine origin** of the sedimentary rock in which the cave is formed.

▲ **Exploration teams also survey the cave** and inventory features such as speleothems, dripping water, living organisms, and cultural features. This meticulous work provides valuable information that is used to create cave maps, assist researchers, and help cave managers protect the cave's fragile resources.

◁ **Park rangers** lead scheduled, guided, cave tours year round. Three different tours follow paved and lighted routes, providing visitors with the opportunity to learn about the cave's geology, features, and management issues.

▲ **W**inter, with its intermittent snowfall and sometimes frigid temperatures, presents a challenge for park wildlife. Bison, with their great strength and heavy winter coats, are well adapted for winter survival. Using their massive heads, they are able to sweep deep snow aside to reach the dormant but nutritious grasses below.

Black Hills Parks and Forests Association

The Black Hills Parks and Forests Association (BHPFA) is a non-profit organization that provides support for Wind Cave National Park, Jewel Cave National Monument, Black Hills National Forest, Custer State Park, and Cleghorn Springs Aquatic Center.

BHPFA operates sales outlets throughout the Black Hills, featuring educational and interpretive publications and other media. Proceeds from sales at Wind Cave National Park are used to publish site-specific materials such as a Wind Cave map, a Wind Cave handbook, trail guides, the park's visitor guide newspaper, and a Junior Ranger booklet. Funding is also provided to develop and improve exhibits, assist in research, and provide cave tours and other interpretive programs.

The Park Today

Wind Cave National Park offers visitors a variety of recreational opportunities. The cave continues to be a major focus of activity with approximately 100,000 visitors annually wandering its maze of passageways on one of five different types of guided tours. The more adventurous travel through the cave by the light of candles as visitors

(Text continues on page 48.)

Wind Cave National Park experiences a milder climate than the surrounding plains and hills to the north. The average annual snowfall is only about two feet. Summer days often record daytime high temperatures in the 90s, and afternoon thunderstorms are common. The highest point in the park is Rankin Ridge at 5,013 feet.

Visitors leave Wind Cave National Park with a collection of images—shaggy bison grazing peacefully on an open prairie, the whistling of the wind through the cave's natural opening, the yipping of coyotes on a still night, the canopy of stars sparkling in a prairie sky, the joyful song of a meadowlark, the spine-tingling bugling of a bull elk, prairie grasses and coneflowers waving in the wind, the scent of the pines after an afternoon thunderstorm, the silent splendor of the cave's winding passageways, and many more too numerous to mention. These images are testimony to the wisdom of preserving this place as a national park for this and future generations to enjoy.

Our national parks are often referred to as national treasures, with individual parks representing jewels in a crown. Although Wind Cave National Park remains less known and less visited than larger national parks such as Yellowstone and Rocky Mountain, those who take the time to know and experience its wonders soon realize that smaller jewels are no less precious.

Visitors admire the beauty of a boxwork ceiling just as they did when the park was established in 1903.

did at the turn of the century or on their hands and knees in a strenuous ranger-led trip. In the cave's outer reaches, explorers continue to pierce the darkness and reveal its ancient, hidden treasures.

Above the cave, the park preserves and protects a remnant of the once vast and undisturbed mixed grass prairie that covered the central portion of the continent. The seemingly endless prairie scene can be startling in its unpretentious beauty and serenity. Numerous wildlife species provide excellent wildlife viewing opportunities for the park's 800,000 annual visitors.

A visit to the park should start at the visitor center, and a nearby picnic area is perfect for a peaceful lunch. The Rankin Ridge fire tower provides a panoramic view of the Black Hills and the Great Plains. Evening campfire programs are offered by rangers nightly during the summer at Elk Mountain Campground.

Most park visitors enjoy the scenery from the comfort of their automobiles while driving the park's two paved highways or two unpaved backcountry roads. An alternative way to experience the park is face-to-face by hiking its 30 miles of trails or traveling cross-country. This experience can lead to a personal appreciation and reverence for this vanishing landscape. The bond between person and place here in Wind Cave National Park becomes stronger with time.

A bumblebee at work on a spotted gayfeather.

Books on National Park areas in "The Story Behind the Scenery" series are: Acadia, Alcatraz Island, Arches, Badlands, Big Bend, Biscayne, Blue Ridge Parkway, Bryce Canyon, Canyon de Chelly, Canyonlands, Cape Cod, Capitol Reef, Channel Islands, Civil War Parks, Colonial, Crater Lake, Death Valley, Denali, Devils Tower, Dinosaur, Everglades, Fort Clatsop, Gettysburg, Glacier, Glen Canyon-Lake Powell, Grand Canyon, Grand Canyon-North Rim, Grand Teton, Great Basin, Great Smoky Mountains, Haleakalā, Hawai'i Volcanoes, Independence, Jewel Cave, Joshua Tree, Lake Mead-Hoover Dam, Lassen Volcanic, Lincoln Parks, Mammoth Cave, Mesa Verde, Mount Rainier, Mount Rushmore, Mount St. Helens, National Park Service, National Seashores, North Cascades, Olympic, Petrified Forest, Rainbow Bridge, Redwood, Rocky Mountain, Scotty's Castle, Sequoia & Kings Canyon, Shenandoah, Statue of Liberty, Theodore Roosevelt, Virgin Islands, Wind Cave, Yellowstone, Yosemite, Zion.

A companion series on National Park areas is the *"in pictures...The Continuing Story."* This series has **Translation Packages**, providing each title with a complete text both in English and, individually, a second language, German, French, or Japanese. Selected titles in both this series and our other books are available in up to 8 languages.

Additional books in "The Story Behind the Scenery" series are: Annapolis, Big Sur, California Gold Country, California Trail, Colorado Plateau, Columbia River Gorge, Fire: A Force of Nature, Grand Circle Adventure, John Wesley Powell, Kauai, Lake Tahoe, Las Vegas, Lewis & Clark, Monument Valley, Mormon Temple Square, Mormon Trail, Mount St. Helens, Nevada's Red Rock Canyon, Nevada's Valley of Fire, Oregon Trail, Oregon Trail Center, Santa Catalina, Santa Fe Trail, Sharks, Sonoran Desert, U.S. Virgin Islands, Water: A Gift of Nature, Whales.

Call (800-626-9673), fax (702-433-3420), write to the address below, Or visit our website at www.kcpublications.com

Published by KC Publications, 3245 E. Patrick Ln., Suite A, Las Vegas, NV 89120.

Inside back cover: *Sunlight ▷ filtered through big bluestem grasses announces the dawn of a new day. Photo by Connie Toops.*

Back cover: *Pronghorn ▷ can thrive only as long as prairies continue to exist. Photo by Jeff Gnass.*

Created, Designed, and Published in the U.S.A.
Printed by Doosan Dong-A Co., Ltd., Seoul, Korea
Color Separations by Kedia/Kwang Yang Sa Co., Ltd.
Paper produced exclusively by Hankuk Paper Mfg. Co., Ltd.